Lily and Zora's Cloud Quest

A Science Story About Clouds and Weather

by Rudo Kashiri

Curious Minds Explorers Publishing

A Curious Minds Explorers picture book

Ages 6 - 10

Lily and Zora's Cloud Quest

A Science Story About Clouds and Weather

Illustrations by Roshanay

Published by Curious Minds Explorers Publishing

Stories That Launch Curiosity

www.curiousmindsexplorerspublishing.com

ISBN: 979-8-9996203-5-4

Printed in the United States of America

This book is dedicated to Erica,

Like clouds, you have shown how gentleness and strength can live side by side, gathering yourself, weathering storms, and giving love freely and wholeheartedly. Watching you grow into the remarkable woman you are, and the devoted, loving mother you are to the twins, has been one of the most beautiful joys under the sky.

"Don't let anyone rob you of your imagination, your creativity, or your curiosity. It's your place in the world; it's your life."
- Dr. Mae Jemison

This book belongs to

____

Every day, clouds float across the sky. They move softly and change shape, like a slow sky parade. If you look closely, clouds can give you clues about the weather.

For a long time, people have looked up at clouds to help them understand the day. By watching and asking questions, they could tell when rain might come, when the sky might clear, and when it was a good time to rest or play.

In this story, Lily and Zora learn to look up together. With their grandma, Ambuya, helping them, they find out that clouds are made from water and air. Clouds are always moving and always changing. They can always teach us something new.

When you watch the clouds, you're not just daydreaming.
You're already thinking like a scientist.

Lily and Zora sat side by side on a rainbow bright rug, legs stretched out and hands behind them, gazing up at the wide blue sky. Above them, clouds drifted past, slowly changing shape.

Brody sat up on the rug too, staring at the clouds like he was imagining shapes of his own.

"Whoa," Zora said, tipping her head. "The clouds are really scooting today!"

Lily felt the breeze brush her face and nodded. "The wind must be giving them a big push."

"Look!" Zora lifted one arm and pointed. "That one's fluffy like cotton candy. Extra puffy!"

"And that one looks like a camel!" Lily laughed. "See? It's walking across the sky with its bumpy back!"

Zora lowered her voice, as if the clouds might hear her. "Clouds are like magic. They can turn into anything."

"And they're never the same," Lily added, her eyes following a cloud as it stretched and curled. "Wispy... puffy... swirly..."

"Nature's art show," Zora grinned.

Lily leaned back and smiled. "I could watch them drift all day."

Behind them, Ambuya's hands slowed in her weaving, as if she knew a good question was coming.

What shapes have you seen in clouds before?

"Ambuya," Lily asked, turning around, "how do clouds stay up there? Why don't they fall?"

Zora leaned in beside her. "And why do they change shapes every day?"

Ambuya set her weaving gently on the ground and smiled.

"Smart questions," she said. "Clouds stay up because air and water are always on the move."

"Warm air rises," she continued and carries invisible water vapor with it."

Lily and Zora tipped their chins back toward the sky, listening closely.

"Higher up, the air is cooler," Ambuya said. "That's where clouds begin their story."

"Some of the vapor turns into teeny-tiny droplets. Lots of tiny droplets and sometimes tiny ice crystals together make a cloud."

"And when the wind changes, clouds can stretch, swirl, and turn into brand-new shapes."

She pointed up. "If we watch carefully," Ambuya said, "clouds can tell us what the sky is doing."

Lily and Zora gazed upward, ready to hunt for sky clues.

"Let's go on a cloud quest," Ambuya said. "Time to collect sky clues."

Do you remember a day when clouds moved fast? What was the weather like?

Ambuya pointed to a small puddle warming in the sun.

"When the Sun heats water," she said, "the water doesn't disappear. It goes up into the air."

Zora leaned closer. "What happens to it once it's in the air?"

Lily tilted her head. "Is it floating up like the clouds?"

Ambuya smiled. "Yes, just like that. It rises as water vapor. That's called evaporation."

"Water vapor is still water," she explained. "It's just so tiny we can't see it."

She added, "In a hot shower, we sometimes see a steamy mist. That mist is **tiny droplets** of water floating in the air, not the invisible water vapor."

Brody sniffed the air, nose twitching, as if he were trying to find the missing water.

Lily laughed as a warm breeze lifted a leaf and carried it up, up, up. "Like vapor!"

Ambuya looked back toward the sky. "And as that water vapor rises, something new begins to happen."

Zora's voice dropped to a whisper. "What is it, Ambuya?"

"The water vapor keeps rising," Ambuya explained, "until it reaches cooler air high in the sky."

Zora's eyes widened. "So the higher it goes, the colder it gets?"

"That's right," Ambuya said.

She lifted her hands. "Let's try it. Warm air rises, so our hands rise too."

Zora and Lily popped up on their toes and stretched their arms high.

"Up, up, up like warm air!" they cheered.

Lily hugged herself and shivered. "Brrr... cool air feels tingly!"

"That cool feeling is what the air is like higher up," Ambuya said.

"That cool air helps water vapor begin changing again," she continued.

"It's like when you breathe out on a cold day and your breath turns into mist you can see."

Zora giggled, squinting at the sun. "Cloud questing makes me thirsty!"

Lily nodded. "Me too. Can we get a cool drink?"

Ambuya smiled. "Yes. Come inside, for a cold glass of water."

Cool Air
Warm
Vapor

Inside, Zora filled a glass with icy water.

As she held it up, tiny drops began to appear on the outside.

She traced one with her finger. "It's traveling! Ambuya, why is my glass getting wet?"

Ambuya smiled. "That water didn't come from inside the glass. It came from the air around it."

Lily's eyes widened. "There's water in the air?"

"Yes," Ambuya said. "The air can hold invisible water vapor. When that warm air touches something cold, like your glass, the water changes into tiny drops you can see."

Zora watched the drops slide and race. "It's like magic."

Ambuya chuckled. "It's called condensation. The same thing happens high in the sky when clouds begin to form."

Tiny droplets sparkled on the glass, like little drops gathering in the sky above them.

Ambuya looked up. "Now imagine billions of tiny drops joining hands, higher and higher."

Lily laughed softly. "Who knew a glass of water could be a science lesson?"

"I want to watch the clouds again!" Lily sprang to her feet.

They stepped outside into the bright sun. Ambuya shaded her eyes, and Lily and Zora did the same, looking up at the sky without staring at the sun.

"Remember the drops on your cold glass?" Ambuya said. "That happens in the sky, too."

"When water vapor cools, it turns into teeny-tiny water droplets."

Zora leaned in, eyes wide. "They're really that tiny?"

Ambuya nodded. "Yes. And those tiny drops need a 'cloud seed,' a teeny speck in the air, to stick to, like dust or pollen floating around."

"The drops gather around that speck," she continued, "then gather around each other, almost like holding hands."

Zora grabbed Lily's hand and spun playfully. "Let's be droplets! Tiny cloud party!"

Ambuya smiled. "Drops join together like a community of friends. And when enough of them come together, they build a cloud we can see floating overhead."

Zora looked up and asked, "Ambuya, how do clouds get their names?"

"Good question," Ambuya said. "Some cloud names tell us their shape. Other names tell us how high they float in the sky."

Lily tilted her head. "So a cloud's name can tell us what it looks like and how high it hangs out."

Ambuya smiled and pointed to the bright blue sky, where big, fluffy clouds drifted by.

"See those?" she asked. "They're cumulus clouds."

"They usually float low in the sky," she continued, "and on many days, they're a sunny sky clue."

Zora grinned. "I love sunny-day clouds."

Brody wagged his tail, as if he agreed.

Lily laughed. "Brody likes them too!"

"When the air is calm, cumulus clouds like to pop up," Ambuya explained. "They don't usually bring rain, just shade and fun shapes to spot."

"So if you spot them, it's a perfect day to play outside and look for animals in the sky."

Zora tilted her head. "When I look at clouds, I like to imagine they're animals."

Lily smiled. "Me too. It's like the sky is telling stories."

Cumulus - Fair Weather

High above the puffy clouds, soft, feathery streaks brushed across the blue sky.

Zora squinted and pointed. "It looks like someone painted the sky with a giant white brush!"

Lily's eyes went wide. "They're so thin, like feathers way, way up!"

Ambuya nodded. "Those are cirrus clouds."

"They form very high in the sky, where the air is cold and windy, so they look wispy, like little brushstrokes."

Zora smiled. "They look light enough to tickle the sky."

Lily tilted her head. "They don't look puffy like the others. Does that mean the sky is changing?"

Ambuya nodded. "Yes. Cirrus clouds can be a clue that a weather change is on the way, maybe tomorrow, maybe the next day."

Lily clasped her hands behind her back. "So the clouds are giving us hints?"

Ambuya smiled. "Exactly. The sky shares clues if we know how to look."

Zora grinned. "So when we see wispy cirrus clouds, we can say, 'Weather clue!'"

Lily nodded. "I'm glad you're teaching us how to read the sky, Ambuya. This cloud quest is the coolest."

Cirrus - Change Coming?

Later, a wide gray cloud slid over the sky like a cozy blanket.

Zora looked up. "The sky looks sleepy."

Lily giggled. "And cozy like it's wearing gray pajamas!"

Ambuya smiled. "Those are stratus clouds, low clouds that spread out like a blanket."

Plip. Plip.

Raindrops began to fall, making tiny splashes in the grass.

Zora squealed and laughed. "Hey! The sky is sprinkling us!"

They dashed into the house and watched the rain tap the window.

Brody happily caught the cool drops on his fur. Then he gave a big shake, spraying water everywhere.

Ambuya pointed to the dark, smooth cloud.

"Stratus clouds hold lots of teeny-tiny water drops," she said. "When those droplets join together and grow heavy enough, they fall as rain."

Ambuya smiled. "That's when it's smart to grab a raincoat or an umbrella."

Lily giggled. "Good thing we brought our jackets!"

Zora watched the rain bounce on the ground. "What if the air is really cold?"

"Then it might fall as snow," Ambuya said. "If raindrops freeze before reaching the ground, they can become sleet.

If liquid drops freeze on cold surfaces, that is freezing rain.

Hail forms in strong storm clouds."

Stratus – Rain could be on the way!

The next morning, Lily and Zora opened the door and stopped. Soft white mist curled around the yard, making everything feel hushed and new.

Lily spun in a slow circle. "Ambuya, are we standing inside a cloud?"

Ambuya stepped beside them and smiled. "Yes. This is called fog, a cloud close to the ground."

Zora reached out her hand. The air felt cool like a tiny, damp kiss on her fingers.

"Fog forms when moist air near the ground cools down," Ambuya said.

"Then tiny water drops gather and float close to Earth."

Lily whispered, then giggled, "It feels like a secret world."

Ambuya nodded. "That's a sky clue: the air is cool and damp."

She lifted a finger toward the brightening sky. "Watch. When the sun climbs higher, the fog may fade away."

Ambuya rubbed her hands together. "Want to see how a cloud can form up close? Let's make one inside a jar."

Lily and Zora squealed and hurried after her. "Yay, an experiment!"

Inside, Ambuya set a clear glass jar on the table and smiled.

Lily and Zora leaned in, eyes sparkling. "What do we do first?"

Ambuya poured hot water into the jar.
"Warm water adds water vapor to the air inside the jar."

"In our jar," she continued, "we can add a tiny helper, one quick spray of hairspray mist."

"Those tiny specks act like cloud seeds, giving water droplets a place to gather."

Next, she covered the jar with a plate and piled ice cubes on top.
"Now warm air meets cold air. Watch."

Inside the jar, a soft, cloudy swirl began to spin.

Lily's eyes widened. "Why is it getting cloudy?"

Ambuya nodded. "The jar follows the same rules as the sky. First, warm air rises. Then cold air cools it down. And tiny droplets form and gather together, just like a cloud."

Zora leaned closer. "We made a cloud!"
Then she blinked. "Is that condensation, like on a cold glass?"

Ambuya smiled. "Yes. Condensation. The warm water made water vapor," she said. "When it cooled down, it turned back into tiny droplets you can see. That cloudy swirl is your cloud."

Why do we need hair spray?
Just like the clouds outside.
Hair Spray
Hot Water

After the fog cleared, they went outside again to look for new sky clues.

Birds sailed high across the sky.

Zora pointed up, eyes wide. "Look how high they're flying! Do you think they're playing tag with the clouds?"

Ambuya chuckled. "Sometimes people notice birds soaring high when the air is calm and the weather is fair."

Lily twirled, gazing up. "And the clouds are so puffy and bright like giant marshmallows!"

Zora grinned. "So, no rain today?"

"Most likely," Ambuya said. "Puffy, bright clouds and high flying birds can be sky clues that the air is calm."

Lily smiled. "It's almost like the sky is whispering little secrets."

Zora giggled. "Then let's listen! What other secrets will the sky tell us today?"

Lily pulled out her Cloud Quest journal and wrote:

"Sky clue for today:
Puffy, bright clouds and birds flying way up high mean the air is calm.
Calm air usually means fair weather!"

As the sun dipped low, the clouds glowed gold, peach, and pink, like a painted goodbye.

Lily leaned against Ambuya. "The clouds are changing again," she whispered.

Zora watched the colors swirl and fade. "It's like the air is stirring them."

Ambuya wrapped her arms around the girls, pulling them close.

"Every time you watch the clouds, you're learning how water, air, and sunlight dance together."

Brody gave one happy bark and thumped his tail in agreement.

Ambuya looked from Lily to Zora, her eyes shining.

"I'm so proud of you both. You kept looking up, asking questions, and learning."

"Don't stop. This is just the beginning. The future is full of possibilities, and I can't wait to see what you discover next."

Lily wiggled with excitement. "We'll keep looking up!" she promised.

Zora squeezed Ambuya's hand and giggled. "And we'll keep asking a zillion questions, because the sky has sooo many secrets!"

What cloud clue do you want to look for next time you're outside?

Later, back in the house, Ambuya opened her tablet and showed Lily and Zora a picture of Earth from far above. The girls leaned closer.

"Look at all those clouds!" Lily said.

Zora's eyes widened. "They're huge! You can see swirls and stripes and giant, puffy patches."

Ambuya smiled. "We can watch clouds from the ground, but scientists study them from space, too."

"From space?" Lily asked.

Ambuya nodded. "Yes. Satellites take pictures of Earth from space. They can watch clouds move over oceans, forests, mountains, and cities."

Zora leaned closer. "So satellites can see the whole sky at once?"

"They can see much more than we can from one spot," Ambuya said. "Some satellites even measure temperature, water vapor, and storm patterns."

Lily looked thoughtful. "So the sky gives clues to scientists on Earth, and to scientists in space."

Ambuya laughed softly. "That's right. Clouds have many stories to tell."

Ambuya's Science Corner

- Clouds are made of tiny droplets of water, and sometimes tiny pieces of ice.
- The Sun warms water on Earth, and some of it rises into the air as an invisible gas called water vapor.
- When the air cools, the water vapor turns back into tiny droplets. This is called condensation.
- The shapes of clouds and where they float in the sky can give us clues about the weather, like "sunny soon" or "rain might be coming."
- Satellites are machines that travel around Earth in space. Some satellites take pictures and collect information to help scientists study clouds, storms, and weather.

Next time you look up at the sky, what clouds will you discover?

Author's Note

When I was young, I learned that some of the best science lessons don't happen in classrooms or laboratories. They happen outside, when we watch the world closely with someone who cares.

Clouds were one of my first teachers.

They drift, change, disappear, and return. They remind us that science isn't about having all the answers right away. It's about noticing, wondering, asking questions, and learning together.

In this story, Ambuya represents the elders, parents, grandparents, teachers, and caregivers who guide children with patience and curiosity. Lily and Zora remind us that children are natural scientists, always observing, questioning, and imagining. And Brody is here to remind us that learning can be joyful, playful, and shared.

My hope is that this book invites you and your child to look up together, talk about what you see, and ask, "What do you think is happening?"

Every question can be the beginning of discovery.

Thank you for exploring the clouds with us.

Rudo Kashiri

Curious Minds Explorers Publishing

Make a Cloud in a Jar

Adult Safety: An adult should handle the warm water and supervise the activity. If using hairspray: spray **one** quick time **away from faces**, close the jar right away, **do not inhale**, and wash hands after.

Steps

1. Adult: Pour warm water into the jar (about ¼ full).
2. Add **one** cloud helper: either a small pinch of salt **or** one quick hairspray spray into the jar (then close it right away).
3. Cover the jar tightly with a plate or lid.
4. Place a few ice cubes on top of the plate/lid.
5. Wait **30 to 60 seconds** and watch for a cloudy swirl inside the jar.

Look and Wonder

- **What you'll see:** a cloudy swirl inside the jar made of tiny water droplets.
- **Why it happens:** the warm water makes **water vapor** (invisible water in the air). The ice cools the air near the top of the jar. When the vapor cools, it turns into tiny droplets. This is **condensation**.

Try Again (Like a Scientist!)

- Try **more ice** vs. **less ice**.
- Try **warmer water** vs. **cooler water** (adult handles).
- Try **with salt** vs. **without salt**.

Question: Which change made your cloud form **faster** or look **thicker**? What do you think caused that?

www.ingramcontent.com/pod-product-compliance
Lightning Source LLC
LaVergne TN
LVHW070900210926
848482LV00016B/979